TABLE OF CONTENTS

The La Brea Tar Pits and Museum is part of the Natural History Family of Museums, which also includes the Natural History Museum of Los Angeles County in Exposition Park, and the William S. Hart Museum in Newhall.

The La Brea Tar Pits have fascinated scientists and visitors alike for more than a century.

This natural scientific wonder in the heart of one of North America's largest cities offers a glimpse into the past, both long-ago and recent. Over time, this land has been forest and savannah, ranch land and oilfield, Mexican land grant

source of asphalt for thousands of years of human use. This continuum of history and prehistory is reflected in the current iconic image of the site, a grouping of Columbian mammoth statues at the Lake Pit, a flooded nineteenth

TAR
PITS

Introduction

In the heart of Los Angeles, just a few miles west of downtown, lies what may be the world's richest deposit of Ice Age fossils.

The Rancho La Brea Tar Pits provide an incredibly complete record of the different plants and animals that have lived in the Los Angeles Basin between 50,000 years ago and today. Ranging in size from towering mammoths to "microfossils," so tiny they may only be seen with the aid of a microscope, this ecologically diverse sample of past life includes diatoms, pollen, seeds, leaves and wood, clam and snail shells, insects, spiders, millipedes, fish, frogs, toads, snakes, lizards, turtles, birds, and mammals—in all, more than six hundred species.

The asphalt-rich sediments in which the fossils were preserved are known as the Rancho La Brea deposits. This name derives from when the area was part of a Mexican land grant called Rancho La Brea ("the tar ranch"). Here *muchos pantanos de brea* (many bogs of tar) were first recorded by the Spanish explorer Gaspar de Portola in 1769. Humans have used the asphalt from this locality since prehistoric times, but its treasure trove of fossils was not recognized until the end of the nineteenth century.

More than one hundred excavations have been made at Rancho La Brea since the early 1900s, and most of the recovered fossils are now housed in the Museum located in Hancock Park at the very center of this unique fossil site.

GEOLOGIC TIME

Evidence from geology, astronomy, physics, chemistry, and biology indicates that Earth is about 4.6 billion years old. In order to discuss events that have taken place during the history of our planet, geologists have divided this span of time into smaller units called eras and epochs. The first 4 billion years saw the formation of the atmosphere, the seas, the continents, and the beginning of life. The Paleozoic Era, from about 542 to 251 million years ago (Ma), was a time of diversification of life in the oceans and colonization of the land by early plants and animals. The Mesozoic Era (also known as the "Age of Reptiles"), from about 251 to 66 Ma, was the time when dinosaurs roamed the earth and when birds, mammals, and flowering plants first appeared. Following a mass extinction at the end of the Mesozoic Era, 66 Ma, mammals and birds diversified and mammals replaced reptiles as the dominant large animals on the planet.

The Cenozoic Era is divided into three periods and seven epochs. The Paleogene Period includes the Paleocene (66 to 56 Ma), Eocene (56 to 34 Ma), and Oligocene (34 to 23 Ma) epochs. The Paleogene was warm and wet; huge snakes, crocodiles, and turtles thrived on land, and sea levels were up to 80 meters (260 feet) higher than they are today. During the Neogene Period, which includes the Miocene (23 to 5.3 Ma) and Pliocene (5.3 to 2.6 Ma) epochs, the earth became cooler and drier; forests converted into grasslands; deciduous trees and large mammals diversified; the Isthmus of Panama formed, joining North and South America; and large oil deposits

GEORGE C. PAGE MUSEUM
LA BREA DISCOVERIES
LA BREA
TAR
PITS
MUSEUM

developed beneath the ocean floors. The most recent Period, the Quaternary, which includes the Pleistocene (2.6 million to 11,700 years ago) and Holocene (11,700 years ago to the present) epochs, was colder still. Huge ice sheets repeatedly covered half of North America; sea levels rose and fell; our species, *Homo sapiens*, evolved in Africa and migrated across the globe; and the majority of large mammals on Earth became extinct.

While the Rancho La Brea Tar Pits have been accumulating and preserving fossils throughout the late Pleistocene and Holocene, the most iconic fossils from this site are of the large mammals that became extinct at the very end of the Pleistocene, between about 15,000 and 11,000 years ago. This interval of time represents a very small and recent part of Earth's history, but an important one for humans; it encompassed the last major episode of global climate warming that the earth experienced, and it was also the time when humans first migrated into North America.

Scientists are now debating whether to add another geologic epoch to the earth's time scale, the Anthropocene ("Age of Humans"). These scientists argue that humans have so significantly impacted Earth's biosphere and geosphere that it is causing permanent, identifiable change in the rocks themselves, change that will be visible millions of years into the future. If the Anthropocene is designated a new global epoch, the Rancho La Brea Tar Pits will preserve a fossil record of this time as well. Human-introduced species, domesticated animals and plants, organisms that are spreading here because of human-caused climate change, and even human trash are getting trapped every day in the tar pits' still-active asphalt seeps, forming a record of life today for future paleontologists to investigate.

GEOLOGIC HISTORY OF THE AREA

Until about 100,000 years ago, the Los Angeles region lay deep beneath the sea. For millions of years, the bodies of single-celled organisms called diatoms accumulated on the ocean floor and got covered by marine sediments. The pressure and heat created by these overlying sediments converted the organic matter into oil. This pressure also pushed the underlying rocks down, a process known as subsidence, creating the feature we now know as the Los Angeles Basin.

When the climate is colder, more water on the earth becomes trapped as ice and sea levels go down. During the last glaciation ("Ice Age") of the Pleistocene, the sea retreated, exposing a flat plain between the edge of the ocean and the Santa Monica Mountains. Stream erosion of the Hollywood Hills resulted in the accumulation of large fan-shaped deposits of river sediment at the mouths of the canyons and extending out over the plain. Over time these alluvial fans extended farther from the mountains and raised the overall level of the plain.

California is known for its earthquakes. Over thousands of years, movements deep within Earth's crust have resulted in the formation of many cracks, faults, and fissures that cut through the layers of gravel, sand, and mud laid down by the streams and into the extensive natural reservoirs of the Salt Lake oilfield 1,000 feet underground. Crude oil seeps slowly upwards along these cracks until it reaches the ground surface. At the surface, the lighter petroleum elements evaporate, leaving behind shallow sticky pools of natural asphalt. Many such asphalt seeps were once visible at Rancho La Brea; some of these can still be seen in Hancock Park today, and new ones frequently emerge in the park and nearby neighborhoods. It is in such asphalt deposits that the Rancho La Brea fossils were preserved.

The seeping of crude oil from the Salt Lake oilfield has been going on for more than 50,000 years. Periodic outpouring of asphalt from fissures and vents continued as the alluvial plain was being built. Some hard, oxidized asphalt lenses that formed long ago at ground level are now found buried beneath layers of sand and gravel that were deposited later, while other vents are still actively seeping asphalt at the surface today. For thousands of years, animals and plants have gotten stuck in shallow surface sheets of viscous asphalt, and over time these asphalt layers have built up into large cone-shaped deposits through continued asphalt seeping and sediment deposition. As the layers of asphalt and sediments build up, they cover the remains of plants and animals stuck in the surface asphalt pools, preserving them.

The "tar pits" visible in the park are the relics of some of the more than one hundred sites excavated for asphalt and for fossils over the last century and a half.

THE ENVIRONMENT OF ICE AGE LOS ANGELES

Geologists, paleontologists, and biologists seek to understand the climate and ecosystems of the past using many different types of information. By comparing ancient sediments with those being formed today, geologists are able to reconstruct the landscape of thousands or millions of years ago. By comparing a fossil organism with its nearest living relatives, biologists and paleontologists are able to understand more about ancient life. Differences between fossil animals and plants and their living representatives can tell us a lot about changes in climate and environmental conditions at different times during Earth's history.

We know from such studies that the Santa Monica Mountains, north of Hancock Park, formed the horizon during the Pleistocene, just like they do today. Streams carried sediment

Look Down

How did the tar get here?

Hancock Park is 1,000 feet above an oil field. The oil oozes all the way up to the surface. Animals and bits of plants get stuck, the animals die, and eventually some remains get covered by asphalt and sediments. La Brea tar has trapped millions of living things over tens of thousands of years and preserved materials like bone, wood, and shell nearly perfectly.

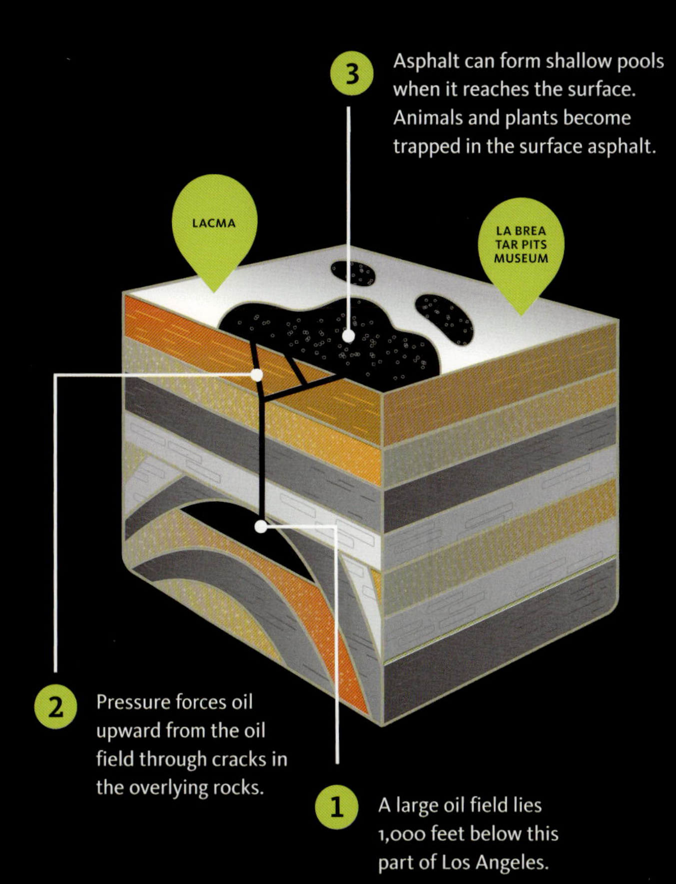

Tar Talk

What's it made of?

La Brea tar is actually asphalt, the lowest grade of crude oil. Asphalt is left over after crude oil's lighter elements, like kerosene, have evaporated.

What stinks?

The "rotten egg" odor is hydrogen sulfide. It comes from the natural breakdown of hydrocarbon molecules in crude oil.

What's bubbling?

The bubbles are methane, a colorless and odorless gas, and hydrogen sulfide. Methane forms with crude oil when marine organisms break down due to heat and pressure.

In the early part of the twentieth century, there were hundreds of oil derricks in a relatively small area between the La Brea Tar Pits and Hollywood. Here, workmen dig in Pit 67, which turned out to be one of the largest and richest deposits.

and debris from the canyons and deposited them on the coastal plain of the Los Angeles Basin. Fossil plants from Rancho La Brea indicate that the late Pleistocene climate was cooler and more humid than our modern climate. Many of the plants found at Rancho La Brea, or their close relatives, occur today in the summer fog belt from San Luis Obispo north to Oregon and on California's Channel Islands.

During the late Pleistocene, therefore, although the shapes of the mountains and valleys would be familiar to the modern residents of Los Angeles, the climate was likely cooler and more humid, the land supported plants now found in colder, wetter areas, and the landscape was inhabited by a variety of different kinds of animals, many of them very big, and some of them unlike anything alive today.

KINDS OF FOSSILS RECOVERED

Early excavators focused their attention on the bones of the larger and more spectacular mammals and birds. The smaller mammals and birds, along with the reptiles, amphibians, fish, invertebrates, and plants, were rarely noticed or collected. Several decades later, workers began to retrieve fossil insects and other small fossils from the sediment, or "matrix," inside some of the large mammal skulls.

In 1969, the staff of the Natural History Museum of Los Angeles County renewed excavations in Pit 91, which they had begun to excavate during their first paleontological dig at Rancho La Brea in 1913–1915. Using modern, more sophisticated techniques, the new excavation sought to document how the fossiliferous deposits in Pit 91 were formed and to recover a greater diversity of fossils. This project has been spectacularly successful in sampling a variety of previously uncollected small animals and plants. Many of Rancho La Brea's "microfossils"—seeds, insects, mollusks, and fish, reptile, amphibian, small

Between 1913 and 1915, hundreds of thousands of fossil bones were excavated from Rancho La Brea. These fossils formed the original collections of the Natural History Museum of Los Angeles County.

Death Trap for Meat Eaters

1 **One fatal footstep**

Thirsty animals step into sticky goo hidden by shallow pools of water, leaves, or dirt.

2 **Struggling makes it worse**

They struggle to get free, tumbling onto their sides. With most of their body stuck in tar, they become hopelessly mired.

3 **Easy pickings for predators**

Dire wolves, saber-toothed cats, coyotes, and other carnivores move in for an easy supper.

4 **Predators fall prey**

Lucky predators tear off exposed limbs and get away, but unlucky ones get stuck and die.

Bones become fossils

After animals died, flesh and soft tissue rotted away, but bone stayed stuck in goo. Seasonal streams washed sand and silt over the bones, fixing them in a sticky mix of asphalt, sediment, and organic remains. Bones kept soaking up tar to become near perfectly preserved.

Ice Age Time Capsules

La Brea's unique asphalt seeps make each pit here a different Ice Age time capsule. This is because the tar oozes up in different places at different times. While one seep stops, another one starts up not far away. The types of plants and animals that get stuck in one seep are often different from those that get stuck in a seep that pops up later. This start-and-stop cycle has occurred many times here over the past 55,000 years.

Fossils form side-by-side, yet twenty thousand years apart

Pools of tar conceal three separate pits (3, 4, and 61/67) dug from 1913 to 1915. Even though these pits are close together, they contained fossils from tens of thousands of years apart. How does this happen?

1 UNDERGROUND AND UNDER PRESSURE

The oil field that underlies La Brea is about 1,000 feet deep. Pressure forces some of the thick oil—asphalt—upward.

2 ASPHALT OOZES UPWARD

As deep underground layers shift, cracks form in the overlying rock layers. Asphalt and methane gas ooze and bubble up through these vents.

3 ACTIVE SEEPS FORM SURFACE POOLS

The asphalt can ooze all the way to the surface to form shallow pools that actively trap animals and plant pieces. Animals decompose and only their bones remain.

4 SEDIMENT BURIES INACTIVE SEEPS

Over many thousands of years, sand and soil wash down from the mountains. This seals off the seeps and protects the fossils trapped inside.

5 GAS BUBBLES CHURN FOSSILS

Methane gas continues to bubble up, churning the bones from different animals and time periods into complex jumbles.

bird, and rodent bones—are best known from this excavation. In all, some 160 species of plants and more than 200 species of invertebrates and 230 species of vertebrates are now known from the Rancho La Brea deposits. Many of these are known only from Pit 91.

In 2006, sixteen new fossiliferous deposits were discovered next to the western end of Hancock Park during the construction of an underground parking structure for the Los Angeles County Museum of Art. The deposits were crated up, intact, in twenty-three large wooden boxes, the largest of which weighed over 120,000 pounds, and relocated to the northern edge of Pit 91. Excavation of Project 23 (named after the number of boxes) has proceeded year-round ever since, using similar techniques to those used in Pit 91 to ensure collection of the rich microfossil record contained in these deposits.

PRESERVATION OF THE FOSSILS

Fossils are evidence of past life. Natural preservation of animals and plants can occur in many ways, but all require isolation from the effects of rapid decay. After an animal dies, its "soft tissues" (e.g., flesh, skin, and hair) usually are scavenged or quickly decay. The harder parts of animals (bones and teeth, shells) tend to decay more slowly and are less likely to be destroyed by carnivores. With very few exceptions, in order for bones, shells, and plants to be preserved as fossils, they must become buried quickly. For this reason, most fossils are found in sediments that accumulated in water (the deposits of ancient rivers, lakes, or oceans) where rapid burial can occur, or other rapid-burial situations like ashfalls from volcanic eruptions. The Rancho La Brea fossils appear to have been preserved by a unique combination of rapid sedimentation and saturation by asphalt.

Asphalt seeps at the Rancho La Brea Tar Pits are most active during warm summer weather, when the asphalt is most fluid, and during times of heavy rains when the high water table helps push the asphalt up to the surface. The resulting shallow puddles are often concealed by a surface coating of dust and fallen leaves. Periodically, an unwary animal becomes trapped by the asphalt. Today the biggest asphalt seeps in Hancock Park are fenced off, but in the past many large animals fell victim to these sticky seeps. One large herbivore stuck in an asphalt seep could lure several predators and scavengers to their fate. The bodies soon decayed, and individual bones became saturated with asphalt and settled at least part way into the mire. These were trampled by other animals that in turn became caught up in the asphalt. During the winter, cool temperatures solidified the asphalt, and fast-running streams deposited a layer of sediment over the exposed bones. The warm weather of early summer reliquefied the seep, and new asphalt oozed to the surface and spread out over the deposit, resetting the trap. Repetition of this annual cycle over thousands of years produced the cone-shaped, bone-filled, asphaltic deposits found at Rancho La Brea.

Very few complete or articulated skeletons have been recovered from Rancho La Brea. Instead, the bones of many individuals of different kinds of mammals and birds are found jumbled up together. The process of this accumulation has still to be established, but trampling by other mammals, fluvial transport, or bubbles of methane gas rising through the seeps may have contributed to the phenomenon. A number of bones found in the deposits display pits and grooves that resulted from rubbing against other bones. This is known as "pit wear", and its cause is still being investigated.

While the process described here is the dominant, and most iconic, form of fossil accumulation at Rancho La Brea, not all fossils at this site are the result of entrapment in asphalt seeps. Some fossils, including the bones of a nearly complete adult Columbian mammoth (nicknamed "Zed"), were washed down in stream channels, and the sediments were secondarily saturated with seeping asphalt, preserving the fossils therein.

The species of mammoth found at the La Brea Tar Pits are Columbian mammoths (*Mammuthus columbi*). Wooly mammoths (*Mammuthus primigenius*), are known from colder climates further north.

Excavation and Preparation Techniques

The position of each fossil in a deposit may provide clues as to how the deposit was formed. Paleontologists use fossil position to investigate such questions as: *was there flowing water at the site when the fossils were deposited?; were the fossils transported prior to being buried?; did an animal die from entrapment in an asphalt seep?; did multiple animals die at the same time?;* and, *what was the relationship between different animals in a deposit?* The documentation gathered during the 1913–1915 excavations divided each fossil-bearing locality into 3-foot-square grids and recorded the depth below the ground surface of each fossil excavated from each grid. This technique recorded the location and depth below ground for most of the mammal and bird bones recovered at that time, but no attempt was made to recover or document other smaller fossils.

DATA ARE RECORDED

Researchers record the fossil's location, depth, and orientation, and take lots of photos before they dig.

FOSSILS ARE CAREFULLY EXCAVATED BY HAND

Dental picks, trowels, small chisels, and brushes are used to remove the fossils.

LARGE SPECIMENS ARE CLEANED BY HAND

A solvent helps researchers remove the asphalt from the fragile fossils.

FOSSILS ARE IDENTIFIED AND CATALOGED

The clean, identified fossil is stored and its data is entered into the master catalog.

THE MATRIX IS PROCESSED FOR MICROFOSSILS

Once cleaned to remove the asphalt, the fossil remains are sorted under magnfications.

Since the reopening of Pit 91 in 1969, excavators at Rancho La Brea have worked to collect all fossils, not just the larger mammals and birds. In addition to location and depth below ground, the orientation of the bones within the deposit is also recorded. Additional information is recorded by means of photographs and sketches of the sediments and the fossils in the ground. The fossils are then carefully excavated with small hand tools: dental picks, trowels, small chisels, and brushes. Each fossil is placed with its excavation data into a separate container for further preparation and cataloging in the laboratory. The sedimentary matrix from which the fossil was recovered is also collected to process for very small fossils (microfossils).

The matrix is placed in screened-bottomed buckets and heated in solvent to remove the asphalt. The cleaned material is a mixture of sand, small pebbles, and microfossils, including seeds and small plant remains, snails, insect parts (mostly beetle fragments), and bones and teeth of small vertebrates. The fossil remains are removed by hand under magnification and identified and cataloged.

Large specimens are cleaned by hand in the laboratory. The sand, clay, and asphalt matrix adhering to the fossils is carefully removed with the aid of a solvent to dissolve the asphalt. Small or fragile specimens may be cleaned in ultrasonic tanks, where high-frequency sound waves help the solvent to remove asphaltic sediment from the specimen. When the cleaning is completed, the specimens are identified and assigned a catalog number, which is entered in a master catalog along with information about the fossil and locality. Each specimen is then incorporated into the collections where it will be available for further research.

RADIOMETRIC DATING

All matter is composed of one or more chemical elements. Carbon, oxygen, nitrogen, and hydrogen are the most common elements in living organisms. Some elements have different forms, known as isotopes, that are defined by the number of protons and neutrons in their nucleus. Some isotopes are stable, but others are not and gradually decay into other more stable forms. The change from unstable to stable isotopes provides the basis for radiometric dating.

The common stable isotope of carbon is called Carbon-12, or ^{12}C, because it has six protons and six neutrons in its nucleus. An unstable isotope of carbon, ^{14}C (with six protons and eight neutrons in its nucleus), is produced in the upper atmosphere by the action of cosmic rays on an isotope of nitrogen (^{14}N). As the ^{14}C spreads through the atmosphere, it combines with oxygen to form carbon dioxide (CO_2), which gets incorporated into plants through photosynthesis, and then into animals who eat those plants. When a plant or animal dies, the ^{14}C in its tissues decays at a fixed and known rate. Every 5,730 years, half of the ^{14}C in a fossil turns back into ^{14}N. So, half of a specimen's original ^{14}C is gone after 5,730 years; after 11,460 years, three-fourths of the total ^{14}C will have decayed; and so on until the levels of ^{14}C in a specimen become undetectable, generally around 50,000 years. Thus, the proportion of ^{14}C in a dead organism compared to the proportion of ^{14}C in the atmosphere can be used to calculate the length of time since the organism died. The radiocarbon dates on animals and plants from Rancho La Brea range from a couple of hundred years to greater than 50,000 years.

Very few complete or articulated skeletons have been recovered from Rancho La Brea. Instead, bones of many individuals are found jumbled up together, in what is commonly referred to as a bone mass.

The Story of Zed

In 2006, a construction crew in Hancock Park found a treasure trove of 16 asphaltic fossil deposits while digging a parking garage for the Los Angeles County Museum of Art.

1 Big crates

These tar pit deposits were collected intact and brought above ground in 23 separate parcels, the largest of which weighed over 120,000 pounds.

2 A plaster present

One of these parcels contained Zed, the most complete skeleton of a Columbian mammoth ever discovered at Rancho La Brea.

3 Preparing the fossils

Over the following years, Zed's skeleton was carefully prepared in the Tar Pits Fossil Lab.

4 Pieces of the puzzle

Radiocarbon dates indicate that Zed lived about 35,000 years ago. Having such a complete skeleton has allowed scientists to learn a lot about Zed's life and death.

Unlike most fossils at Rancho La Brea, Zed didn't get caught in an asphalt seep.

Instead, his body was washed into a stream, and his bones were preserved by asphalt after burial. Zed's bones were found with a lot of other material from the stream, like fish and freshwater snails.

A telling tooth

From studying the layers of Zed's tusk, scientists have learned not just about how he lived, but how he may have died. Like modern male elephants, Zed stopped eating during the musth season, but one year he never recovered. Researchers think that he may have died of soft-tissue injuries sustained during combat for females.

It is fortunate that the fossils from Rancho La Brea are young enough that their tissues can be directly dated by radiocarbon. Materials beyond the limits of radiocarbon dating have to be dated using elements that are less abundant than carbon in body tissues. The decay of radiometric potassium to argon is the most frequently used method. Radiometric dates for fossils older than 50,000 years are usually estimated from potassium-rich rocks (such as volcanic lavas and ashes) occurring at the localities in which fossils are found.

RECONSTRUCTING EXTINCT ANIMALS

The Rancho La Brea fauna is composed of both extinct and extant (still living) species. Some of the extant species are still found in the Southern California region; others no longer occur in the area. Many of the extinct species were larger than their modern equivalents. Comparing the extinct forms with their closest living relatives provides information on physiology and behavior. Some extinct species have no living relative, and reconstructing the physical appearance of these animals requires knowledge of the structure and function of their bones.

The initial stage of any reconstruction is to identify each of the animal's bones and to place the bones in their proper skeletal position. After the skeleton is assembled, the soft tissues can be modeled. Muscles sometimes leave distinct scars where they were attached to bones, and these can be used to estimate the size and proportions of the muscles. Hair, fur, and skin are rarely found in the fossil record, and so the external appearance (color and type of hair or feathers) of the extinct animal is usually based on those of modern relatives and anatomically similar animals.

MIGRATIONS BETWEEN CONTINENTS

The composition of the North American fauna has changed greatly through time. Part of the change was due to the evolution of species on this continent. Other changes in the composition of the North American fauna resulted as animals migrated here from other continents. During the later part of the Cenozoic Era, faunal exchanges took place through two corridors.

The Bering Strait, now covered by a shallow sea separating Alaska and Siberia, was the pathway for interchange between North America and Eurasia for much of the last 65 million years, and this land bridge remained largely ice free during the Pleistocene. Through this route, species that evolved in North America, like dogs, camels, and horses, migrated to Eurasia, and Eurasian mammals, like mammoths, cats, and deer, entered North America.

The other region that exchanged fauna with North America was South America. For tens of millions of years, nearly the entire Cenozoic, South America was an island, isolated from North America and all other continents. The first exchange of animals between North and South America occurred about 9 million years ago across a volcanic island chain. Later, about 3 million years ago, the Isthmus of Panama formed, linking the two continents, and the pace of faunal exchange increased dramatically. The ancestors of the ground sloths represented in the Rancho La Brea fauna traveled up from South America. At the same time, rabbits, rodents, dogs, cats, proboscideans, horses, camels, and deer entered South America from the north.

Based on changes in the North American fauna resulting from a combination of evolution and migration, paleontologists have been able to identify a sequence of successive mammalian faunas (groups of mammal species). These faunas have been used to characterize the North American Land Mammal Ages.

BERING
LAND BRIDGE
ON THE MOVE
Cats first migrated from Eurasia to North America, and then continued on to South America.
Eurasia to
North America
Mammoths, Mastodons,
Gomphotheres, Bison, and Cats
NORTH AMERICA
North America
to Eurasia
Dogs, Camels, and Horses
PANAMA
LAND BRIDGE
North America
to South America
Camels, Deer, Rabbits,
Gomphotheres, Dogs, and Cats
SOUTH AMERICA
MISSED CONNECTIONS
The continents of North America and South America weren't connected until 2.8 million years ago.
South America
to North America
Ground Sloths, Porcupines,
Armadillos, and Capybaras

Rancho La Brea and Our Changing Climate

Few things are more critical to the future of life on our planet, and to the survival and quality of life of our own species, than the Earth's climate.

The environment of any given place on Earth is determined by several factors, including distance from the equator, geographic setting (elevation, distance from the sea, etc.), and prevailing weather patterns. In addition, many factors can cause climate to change on a global scale. For example, volcanism, changes in the arrangement of continents and oceans, the uplift and erosion of mountain ranges, and periodic changes in the distance between the earth and the sun have been responsible for significant changes in climate throughout Earth's history.

During the time that the tar pits at Rancho La Brea were collecting fossils, North America was experiencing significant climate changes. As the world moved through and then out of the last ice age of the Pleistocene, temperature, rainfall, and vegetation changed across the landscape. During much of the time that Pit 4 was active, between 24,000–14,000 years ago, half of North America was covered by ice sheets. In contrast, many of the fossils found in Pits 10 and 61–67 date to less than 10,000 years ago, when California's climate was similar to what we see today. Numerous studies at Rancho La Brea document the changes in Los Angeles' flora and fauna through these transitions, from changes in body size and diet that appear correlated with temperature changes, to the past existence in the Los Angeles Basin of species now found only far to the north. These studies tell stories of the resilience, adaptability, and vulnerability of species throughout time.

Today our climate is changing again, this time from human causes, primarily the release of greenhouse gases into our atmosphere through the burning of fossil fuels. As greenhouse gas levels and global temperatures continue to rise, some scientists are looking to the fossil record to try to predict how plants and animals might respond to these changes, based on how they have responded in the past. No record on earth is as apt for this type of research as Rancho La Brea.

Is climate science buried in asphalt?

FOSSIL	WHAT THIS TELLS US
Rainbow trout bones	Today, rainbow trout live in year-round streams. Finding their fossils in Pit 91 tells us that streams once flowed here all year long.
Monterey cyprus cones	Today, Monterey cypress trees live only where it is cool and rainy. Because they once lived here, winters then were probably colder and wetter.
Jaw of a fence lizard	Fence lizards are one of the many small animals that lived here up to 44,000 years ago. Unlike the saber-toothed cat and some of the other megafauna, most of these small animals survive today.

65 Million Years Ago Nonavian dinosaur extinction

20 Million Years Ago Formation of L.A. Basin oil deposits

2.6 Million Years Ago Pleistocene Epoch begins

100,000 Years Ago
Los Angeles Basin emerges from ocean

240,000 Years Ago
Bison in present-day continental US, Rancholabrean Land Mammal Age

RANCHO LA BREA: CHANGE THROUGH TIME AND TEMPERATURE

The asphalt seeps at Rancho La Brea became active during the very end of the Pleistocene epoch, which spanned from 2.6 million years ago to 11,700 years ago. During the Pleistocene, earth underwent repeated, extreme climate fluctuations, known as Ice Ages. These resulted in land area being lost and re-exposed many times, along with the constant reorganization of plant communities, all of which had significant impacts on animal populations. The Pleistocene also saw the evolution and global migration of modern humans.

Temperature Change Over Time*

50,000 Years Ago
Beginning of Rancho La Brea asphalt deposits

55,000 Years Ago
50,000 Years Ago
45,000 Years Ago
40,000 Years Ago
35,000 Years Ago
30,000 Years Ago

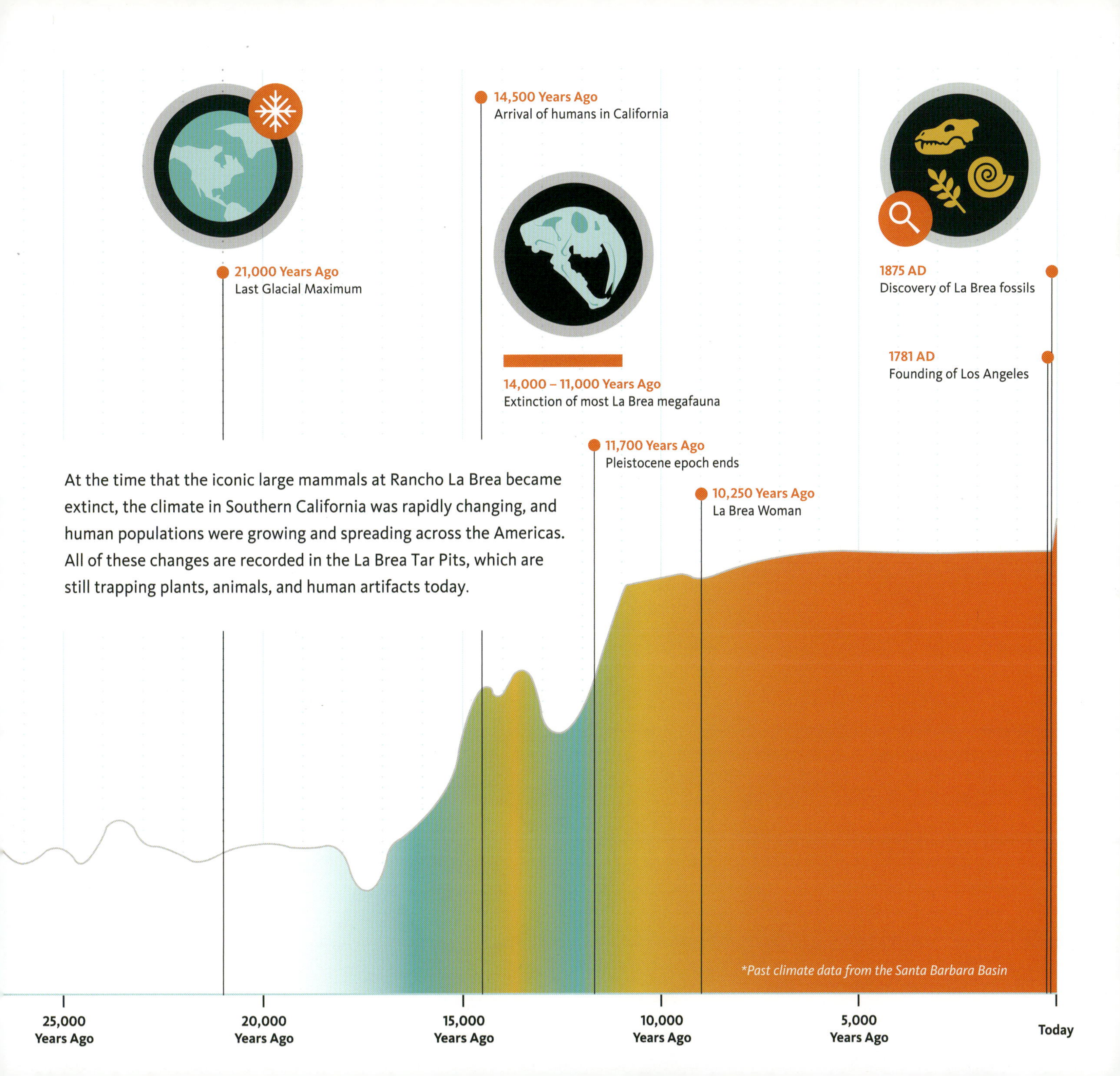
14,500 Years Ago
Arrival of humans in California
21,000 Years Ago
Last Glacial Maximum
1875 AD
Discovery of La Brea fossils
1781 AD
Founding of Los Angeles
14,000 – 11,000 Years Ago
Extinction of most La Brea megafauna
11,700 Years Ago
Pleistocene epoch ends
At the time that the iconic large mammals at Rancho La Brea became extinct, the climate in Southern California was rapidly changing, and human populations were growing and spreading across the Americas. All of these changes are recorded in the La Brea Tar Pits, which are still trapping plants, animals, and human artifacts today.
10,250 Years Ago
La Brea Woman
*Past climate data from the Santa Barbara Basin
25,000 Years Ago
20,000 Years Ago
15,000 Years Ago
10,000 Years Ago
5,000 Years Ago
Today

Imagining Ice Age Los Angeles

UP IN THE AIR

Imagine looking up in the Ice Age sky. You'd see oak and juniper trees swaying in the wind, songbirds and birds of prey flying overhead, and insects buzzing past. Bats appear at dusk, swooping to catch bugs for dinner in the falling light.

SONGBIRD

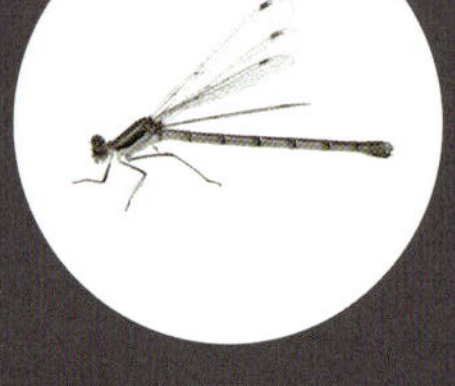

DAMSELFLY

BEETLE

ON THE GROUND

Imagine looking all around at ground level. You'd see acres of sagebrush, wild lilac, and other shrubs sheltering snakes, rabbits, and lots of rodents. Quails coo and rustle in the underbrush, looking for seeds to eat. Coyotes aren't far from sight.

HORNED LIZARD

RATTLESNAKE

SKINK

UNDERGROUND AND UNDERWATER

Imagine listening to the year-round stream that ran right past here. Fish dart in its water and frogs croak on its bank. This water and the soil underfoot nurture oak and juniper trees. The streams and woodlands of Ice Age L.A. help support abundant wildlife.

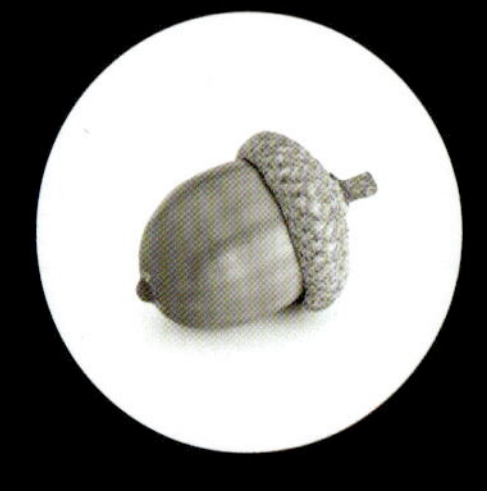

OAK TREE

VOLE

CYPRESS

Evidence from La Brea's small fossils tells us what plants and wildlife lived here during the past 50,000 years. Small fossils found here suggest that this area had juniper woodlands and year-round streams during the Ice Age, when L.A. was cooler and wetter than it is now.

BAT

COTTONTAIL RABBIT

QUAIL

SHREW

GROUND SQUIRREL

CHIPMUNK

DEER MOUSE

ANT

JUNIPER

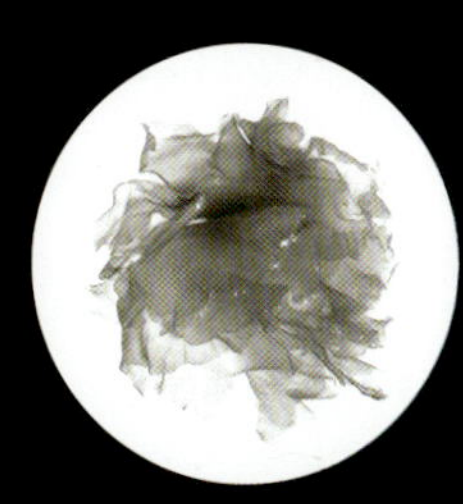

GREEN ALGAE

FROG

STICKLEBACK FISH

Mammals

Over one million mammal fossils, representing at least fifty-nine different species, have been recovered from the asphalt deposits.

Some species, particularly those of the larger carnivores, are represented by thousands of bones and teeth. Other species, such as tapirs and bats, are known only from a handful of specimens. The early collectors focused chiefly on the larger mammals. Our knowledge of the smaller forms (insectivores, rodents, etc.) is still far from complete.

MAMMOTHS AND MASTODONS

Mammoths originated in Africa about 5 million years ago and entered North America around 1.8 million years ago. While most people imagine long-haired woolly mammoths when they think of these animals, the species found at Rancho La Brea is the Columbian mammoth, *Mammuthus columbi*, that is found throughout much of the United States and into Mexico and Central America. Columbian mammoths did not need or have heavy coats, but instead had bare skin with sparse hairs like elephants today. Also like modern elephants, mammoths had flat, high-crowned ("hypsodont") teeth adapted for grinding up silica- and grit-filled grasses. Columbian mammoths occupied North America at the same time as the woolly mammoth,

MAMMAL MANIA!

We've found an incredible number of large mammals in the tar pits, including at least:

2200 Saber-toothed cats

4000 Dire wolves

36 Columbian mammoths

60 Harlan's ground sloths

80 American lions

30 Short-faced bears

300 Ancient bison

220 Western horses

36 Yesterday's camels

...AND COUNTING!

Mammuthus primigenius, which was restricted to the colder high latitudes, and the dwarf mammoth, *Mammuthus exilis,* known from the California Channel Islands.

Columbian mammoths have been found at a number of localities throughout the Los Angeles basin and Southern California. Remains of more than thirty-five individuals have been recovered from Rancho La Brea, most of them from Pit 9 which was excavated in 1913. A relatively complete skeleton of an adult male mammoth, including both tusks, was recovered in 2006 during the excavation of an underground parking structure for the Los Angeles County Museum of Art, which sits adjacent to the La Brea Tar Pits and Museum in Hancock Park. The other mammoth specimens found at Rancho La Brea comprise isolated bones and teeth.

American mastodon, *Mammut americanum* (left) and Columbian mammoth, *Mammuthus columbi*

The Rancho La Brea mammoths were about the size of an African elephant; adult males would have stood 12 to 13 feet tall at the shoulder and weighed 8,000 to 10,000 pounds. Although the mammoths from Pit 9 include both very young and very elderly individuals, most were young adult males (30 to 40 years old at the time of their death), an age estimate based on the state of wear of the teeth and the degree of fusion of the femurs (thigh bones).

Why mammoths were found at so few of the asphalt seeps is still a mystery. Perhaps these were the only localities where conditions were such that mammoths could become trapped by the seeping asphalt. Or perhaps mammoths were present only in the Los Angeles Basin for a short span of time and these were the only seeps active at that time. Unfortunately, no reliable bone dates for the mammoth fossils have yet been obtained, but dates from fossil wood excavated from Pit 9 suggest an age of up to 55,000 years.

Also originating in Africa, the more basal proboscideans known as gomphotheres and mastodons arrived in North America about 10 million years before mammoths. Some gomphotheres had upper and lower tusks, but mastodons, like mammoths, had only upper tusks. Mastodons were smaller than mammoths and had low-crowned teeth adapted for browsing on leaves and shrubs. All of the permanent mastodon teeth could be present in the jaw at the same time, whereas mammoths, like elephants, never have more than two teeth at a time erupted

While nearly all of the mounted skeletons in the Museum are made up of real bones excavated from the Tar Pits, none represent a single animal. Because bones of trapped animals got scavenged and jumbled up, these mounts, of an adult and baby mastodon, are composites, containing bones of many individuals, often from two or more different pits thousands of years apart in age.

in each jaw. The last surviving mastodon species, the American mastodon, *Mammut americanum*, was widespread in North America during the late Pleistocene and has been recovered from many of the Rancho La Brea excavations.

GROUND SLOTHS

Ground sloths were large, strange-looking mammals related to the smaller, present-day tree sloths of Central and South America. They originated in South America but began to migrate to North America when the two continents became close about 9 million years ago. Two species of ground sloths are common in the Rancho La Brea deposits and a third is represented by a few

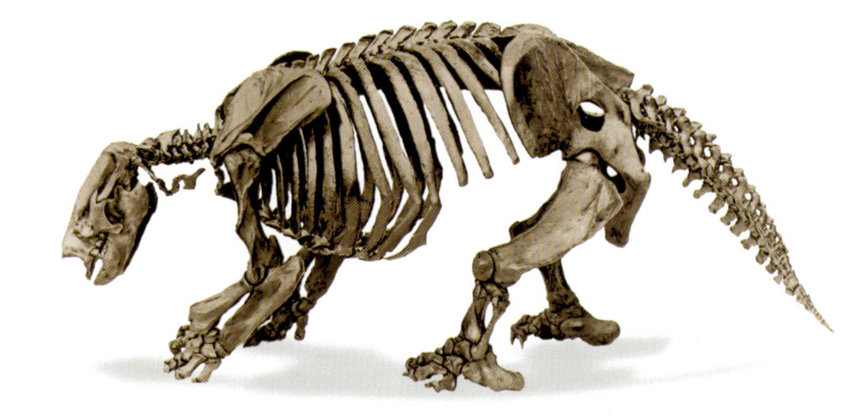

Harlan's ground sloth, *Paramylodon harlani*

specimens. All are characterized by features setting them apart from most other mammals: ever-growing teeth that lack enamel, a big flange on the cheek for the attachment of powerful chewing muscles, an additional set of (sternal) ribs, and modification of the pelvis and vertebrae to allow the animal to sit upright.

Harlan's ground sloth, *Paramylodon harlani*, stood a little over 6 feet tall when upright and weighed about 3,500 pounds. Its simple, flat grinding teeth indicate that it preferred a diet of grass, although it may also have fed on tubers, leaves, and twigs. It had very long claws that may have served for both feeding and self-protection. It also had nodules of bone, called dermal ossicles, embedded in the deeper layers of the skin that served as a kind of armor against attacks by predators.

The smaller Shasta ground sloth, *Nothrotheriops shastensis*, had a tubular-shaped snout and fewer teeth than its larger relative. The structure of its teeth suggests that it was a browser, feeding on leaves from shrubs or low-hanging tree branches. A second, larger browsing ground sloth, Jefferson's ground sloth, *Megalonyx jeffersonii*, is rare at Rancho La Brea and represented by only a few specimens. Neither of the browsing species had the bony ossicles found in *Paramylodon harlani*.

Ground sloths

were tough enough to shake off large predators, but couldn't get loose from the tar.

Best defense?
Razor sharp claws and bony armor hidden in its skin

Closest living relative?
Tree sloth

Favorite food?
Grass, leaves, tree roots, and twigs

HORSES AND TAPIRS

Biologists divide animals with hooves, or "ungulates," into two groups: those that have even numbers of toes (our "cloven hoofed" animals of today—cows, pigs, camels, deer, etc.) and those that have odd numbers of toes. This latter group, called "perissodactyls," comprises three living types of animals: horses, tapirs, and rhinos. Two of these, horses and tapirs, are found at Rancho La Brea.

Horses originated and underwent most of their evolution in North America but migrated out into the rest of the world on several occasions. Horses evolved very tall teeth for feeding on grass and long legs for running swiftly in open country. Most of the horse bones recovered from the deposits belong to the extinct western horse, *Equus occidentalis*. Standing almost 5 feet high at the shoulders, this animal was as tall as a modern Arabian horse but more heavily built. The western horse had high-crowned cheek teeth with a complex pattern of resistant enamel. These provided an efficient grinding surface suited for processing abrasive fodder.

A second smaller and more lightly built species, *Equus conversidens*, is also represented by a few specimens. These two species were some of the last horses native to North America, all of which became extinct about 11,000 years ago. The wild horses found in some parts of America today are the descendants of domestic horses that were brought here by Spanish colonists.

The tapir, a three-toed distant relative of the horse, is adapted for living in the dense vegetation surrounding swamps. Tapirs have low-crowned teeth for browsing on leaves and other

Western horse, *Equus occidentalis*

soft plant food, and their nose is modified into a short proboscis or "trunk," which is used to grasp and push food into the mouth. Like horses, they were formerly native to North America, Europe, and Asia but disappeared from North America during the Pleistocene and are found today only in South America and Malaysia.

Although fossil tapir bones have been found at several sites in Southern California, the only evidence for tapirs at the tar pits consists of two toe bones and a foot bone excavated from the tar pits and a jaw that was recovered during construction a block away from Hancock Park.

Tapir, *Tapirus* sp.

PECCARIES

Zoologists divide the living swine into two families, the Suidae (pigs) of the Old World and Tayassuidae (peccaries), now restricted to the New World. Peccaries occur today from southwestern United States to Patagonia, but they were formerly more widely distributed and occurred as far afield as Kenya and South Africa. They differ from pigs in small but important details of the skull and teeth.

Peccary remains in the tar deposits include a partial skull and several foot bones of the flat-headed peccary, *Platygonus compressus*. This animal ranged widely through the Americas during the Pleistocene Epoch but went extinct along with most large mammals around 11,000 years ago. A related species, *Catagonus wagneri*, was thought to have gone extinct then as well, but was discovered in the 1970s living in the Chaco region of Argentina, Bolivia, and Paraguay. The Rancho La Brea peccary achieved the size of the European wild boar and was somewhat larger than its living North American relatives.

Flat-headed peccary, *Platygonus compressus*

Yesterday's camels

were fast runners, but not with their feet stuck in tar.

Favorite food?
Grass, leaves, and pretty much anything green and edible

Closest living relative?
Although they looked somewhat like today's llamas, new research suggests they were more closely related to today's camels.

Best defense?
Speed! Scientists think these guys could sprint at 40 miles per hour, much like their modern camel cousins.

CAMELS

The earliest camels are from North American sites that have been dated at about 45 million years old. They subsequently migrated to Eurasia, Africa, and South America. Today, the camel family is represented by the two-humped Bactrian camel of Asia, the single-humped dromedary of North Africa and the Middle East, and the llamas of South America. The camel family at Rancho La Brea is represented by two species, both now extinct. Sometimes known as "yesterday's camel," *Camelops hesternus* was the larger and more commonly preserved species. It stood more than 7 feet tall at the shoulder and was of similar build to the living Bactrian camel. *Camelops* was formerly thought to be closely related to the living llamas from South America, but recent DNA evidence suggests that it is probably ancestral to Old World camels.

Yesterday's camel,
Camelops hesternus

Also represented by a few limb bones and vertebrae is the extinct long-headed llama, *Hemiauchenia macrocephala*. Like today's South American llamas, alpacas, and vicuñas, this creature had a long neck and slender legs and probably lacked a camel-like hump. The species of *Hemiauchenia* from Rancho La Brea was about one-third larger than any living llama. *Hemiauchenia* is also found in Pleistocene deposits in South America.

DEER

Although deer are now the most abundant of the large native animals of California, they are represented at Rancho La Brea by only a few individuals of one species, the mule deer, *Odocoileus hemionus*. This species has low-crowned cheek teeth adapted for chewing the leafy vegetation of bushes and trees and limbs modified for running and jumping through rough terrain.

Deer migrated to North America from Eurasia about 3 million years ago, during the late Pliocene. The remains of deer have been found at a number of other Pleistocene localities in California, and their rarity at Rancho La Brea suggests that the area may not have been a suitable deer habitat, or that competition from larger herbivores might have restricted their numbers.

PRONGHORNS

Pronghorns are often called antelopes but are only distantly related to the antelope and cattle of the Old World. The horns of the pronghorn are retained throughout the life of the animal but are covered by a horny sheath that is shed annually. Members of the pronghorn family (Antilocapridae) may, in this way, be distinguished from other horned mammals, such as the true antelopes and cattle (whose unbranched horns are not shed), giraffes (whose "horns" are covered with skin), and deer (whose branched antlers are shed annually).

Pronghorns are also the second fastest land animal on earth, after the cheetah. No predator living in America today can come close to catching a pronghorn in a chase. The pronghorn's incredible speed almost certainly evolved in response to faster, now-extinct megafauna, like the American cheetah.

Although represented today by a single species found only in western North America, pronghorns were more diverse and more widely distributed during the Pleistocene. Two species of pronghorn have been recognized in the Rancho La Brea deposits. The larger species was very similar in size and appearance to the modern pronghorn, *Antilocapra americana*, but is known from only a few specimens. Much more abundant are the remains of a dwarf pronghorn, *Capromeryx minor*, which stood less than 2 feet tall at the shoulder and is different from *Antilocapra* in having horns with two distinct prongs arising from their base.

Dwarf pronghorn, *Capromeryx minor*

Before large specimens like this bison jawbone are cleaned by hand, they are soaked in solvent to remove the majority of the asphalt.

BISON

Bison belong to the cattle family (Bovidae) and are first recognized in the fossil record of Eurasia nearly 2 million years ago. Their arrival in North America from Beringia, about 240,000 years ago, marks the beginning of the part of the late Pleistocene Epoch known as the Rancholabrean Land Mammal Age.

Bison are the most common large herbivores in the Rancho La Brea fauna. The ancient bison, *Bison antiquus*, is represented by more than three hundred individuals. This species was similar to but a little larger than today's plains bison, *Bison bison*. Although bison are specialized to feed on grass, food particles caught in their teeth indicate that the Rancho La Brea bison also ate the leaves of trees and bushes.

Long-horned bison, *Bison latifrons*

On the basis of their tooth eruption, the *Bison antiquus* from Rancho La Brea fall into distinctive age categories of 2 to 4 months, 14 to 16 months, 26 to 28 months, 38 to 40 months, and adults. This suggests that these bison weren't present at Rancho La Brea year round but migrated into the Los Angeles Basin at a specific time each year. It seems likely that the bison herds were present in the vicinity of Rancho La Brea during the late spring when there was a lot of grass after the winter rains. In contrast, the fossil horses have a more continuous age distribution and were in the Los Angeles Basin year-round.

A second, even larger, species of bison is represented by a few individuals: the long-horned bison, *Bison latifrons*. It had a 6-foot spread of horns, stood over 7 feet high at the shoulder and weighed over 2,000 pounds. The long-horned bison has only been found in the older deposits from Rancho La Brea.

MUSK OXEN

Although musk oxen have not been found in the Rancho La Brea deposits in Hancock Park, a site of similar age located only ten blocks from the park has yielded remains of a musk ox, *Euceratherium*. This group of bovids is more closely related to sheep than to cattle. Musk oxen are today restricted to Greenland and the Arctic region of North America but formerly ranged as far afield as South Africa.

Male or Female?

It is not always easy to tell the sex of fossil mammals. Males are often larger than females but size isn't always reliable because of so much variation between individuals. Horns and tusks are sometimes only present in males; if present in both sexes, those of the males are usually larger.

Male fossil skull
Length: 34"
Width: 20"

Ancient bison, *Bison antiquus*

Female fossil skull
Length: 25"
Width: 19"

DOGS

The dog family (Canidae) is one of the oldest groups of modern carnivores; earliest representatives are from the Eocene Epoch of North America. Over thirty genera of doglike fossil mammals occurred in North America. Most of them were adapted for running swiftly in pursuit of their prey.

Representatives of the dog family are the most abundant animals from the Rancho La Brea deposits. The most common species is the dire wolf, *Canis dirus*, known from the bones of more than four thousand individuals. It is likely that packs of dire wolves attempted to feed on animals trapped in the asphalt and while doing so became mired themselves.

The dire wolf was similar in size to the closely related gray wolf, *Canis lupus*, which is also found at this site. The dire wolf had a massive head with strong jaws, large teeth, and legs that were proportionately shorter than those of the gray wolf. Dire wolf bones from the tar pits include many examples of healed injuries such as crushed paws and fractures of the head, limbs, and trunk. These injuries were likely incurred as the dire wolves hunted large prey, and scientists think their social structure allowed these injured animals to survive although unable to hunt by themselves.

Coyotes, *Canis latrans*, are the third most abundant mammal species from the Rancho La Brea deposits. The fossil coyotes are slightly larger than those found in the Los Angeles area today but were significantly smaller than the dire wolves. Their numbers suggest that coyotes frequently visited the tar seeps in search of trapped animals. However, the chemical composition of their bones and teeth suggest that the coyotes were also eating insects and plant material.

Fossil specimens of the living gray fox, *Urocyon cinereoargenteus*, have been recovered in small numbers. The mainland gray foxes gave rise to the Dwarf Island Fox in the California Channel Islands. Two different types of domestic dog have also been recovered from the younger accumulations.

Dire wolf, *Canis dirus*

Coyote, *Canis latrans*

TOUGH PUPPIES

Dire wolves are the most common large animal found at the La Brea Tar Pits, probably because they hunted, and got stuck, in large packs. The thousands of dire wolf specimens uncovered here allow scientists to study how these animals lived, how they responded to environmental changes, and even why they became extinct.

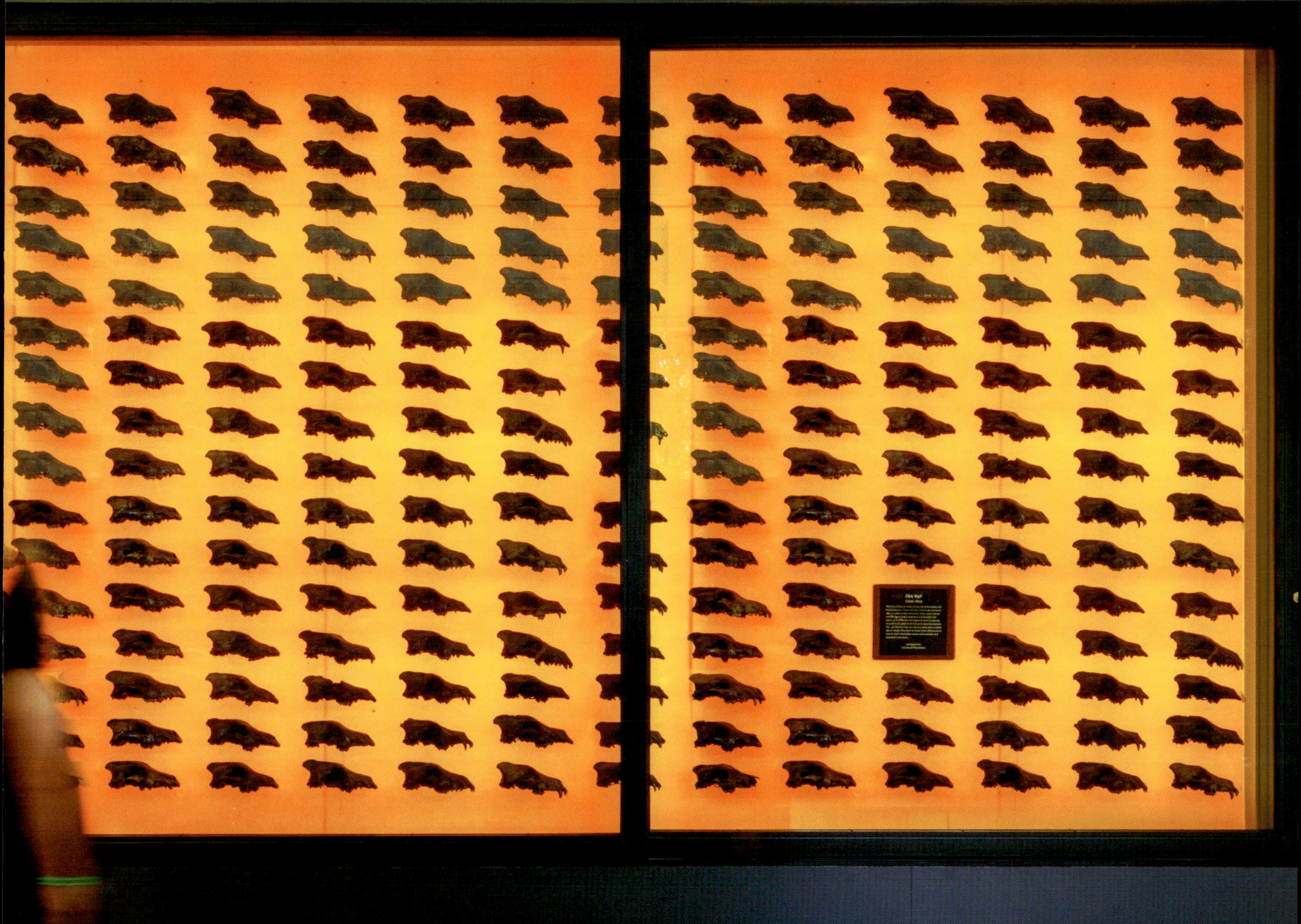

Over four hundred dire wolf skulls, from Rancho La Brea, can be seen in the Museum's dire wolf wall. The tar pits' ability to capture large numbers of individuals is part of what makes Rancho La Brea such an important resource for scientists.

BEARS

Bears evolved from doglike ancestors during the Oligocene Epoch. Originally they had a carnivorous diet, but they subsequently adopted a mixed and largely herbivorous diet. Although the earliest bears are from Europe, they are known to have occurred in North America from the beginning of the Miocene Epoch.

Three different types of bears have been found in the Rancho La Brea deposits. The most common is the short-faced bear, *Arctodus simus*. This is a large extinct species that was limited to North and South America. It had unusually long legs and was taller than the Kodiak bear from Alaska, although it was less heavily built. Most bears are opportunistic feeders, and their teeth have been modified to cope with a diet that includes both meat and vegetation. The short-faced bear, however, more closely retained the primitive carnivore tooth pattern, and studies of its bone chemistry suggest that it ate more meat than its living counterparts. The only living close relative of the short-faced bear is the spectacled bear, *Tremarctos ornatus*, from South America. Two other bears represented at Rancho La Brea are the black bear, *Ursus americanus*, and the grizzly bear, *Ursus arctos*. The black bear is present, though uncommon, throughout the Rancho La Brea sequence; the grizzly bear is known only from the younger accumulations.

Short-faced bear, *Arctodus simus*

SHORT-FACED BEARS, GRIZZLY BEARS, AND BLACK BEARS

are all found at Rancho La Brea. This fossil skull is from a short-faced bear. It is the most common, the largest, and the only one to have gone extinct.

WEASELS, SKUNKS, AND RACCOONS

The smaller carnivores from Rancho La Brea are all living today. The weasel, *Mustela frenata*, was the most common and is represented by more than fifty skulls. Another member of the weasel family (Mustelidae) known from the asphalt is the badger, *Taxidea taxus*. All appear to have been somewhat larger than their living counterparts.

The weasel family has a fossil record of over 30 million years, extending back into the Oligocene Epoch. However, as is often the case with small forest-dwelling animals, the few known fossil remains are very fragmentary. Badgers are known from the Miocene Epoch onward.

Two species of skunks are present in Rancho La Brea, the striped skunk, *Mephitis mephitis*, and the spotted skunk, *Spilogale gracilis*. Formerly classified in the weasel family (Mustelidae), skunks are now recognized as a distinct family of its own (Mephitidae) that has its own evolutionary history going back to the Miocene.

Two members of the raccoon family (Procyonidae), the ring-tailed cat, *Bassariscus astutus*, and the raccoon, *Procyon lotor*, are known from the Rancho La Brea deposits. Both are very rare. This group appears to have originated in Europe during the Miocene Epoch but radiated into North America shortly thereafter and to South America by the late Miocene.

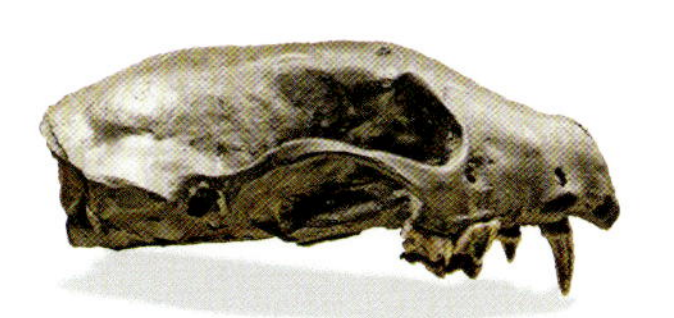

Fossil skulls
Striped skunk, *Mephitis mephitis*

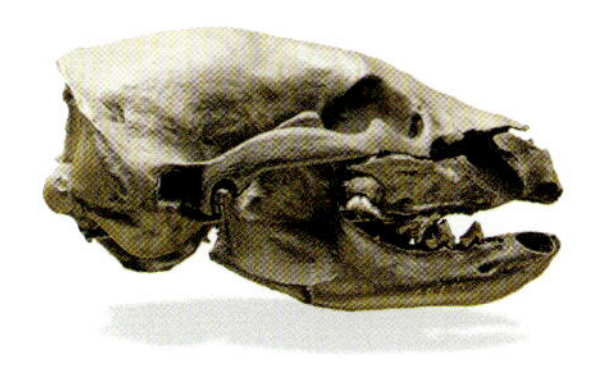

Badger, *Taxidea taxus*

The saber-toothed cat is the most iconic species at Rancho La Brea. Saber teeth have evolved many times in animals, from cats to deer, marsupials, and even fish.

CATS

Members of the cat family (Felidae) first appear in the fossil record of Europe about 25 million years ago. The largest cat from the Rancho La Brea asphalt deposits is the American lion, *Panthera atrox*, which along with the Pleistocene cave lion of Europe is the largest cat that has ever lived. Female *P. atrox* were the size of modern African lions, but the males were about 25 percent larger. The skeleton and teeth of this extinct species strongly resemble those of modern African and Asian lions, and genetic evidence indicates the *P. atrox* is closely related to modern and other extinct lion species, although it also has some features similar to living jaguars.

Pumas are represented by only three individuals, perhaps because, like today, they rarely ventured out of the nearby hills. Bones of the bobcat, *Lynx rufus*, from the Rancho La Brea deposits resembles the species still found in California. Bobcat bones are also not common at Rancho La Brea, but vary sufficiently in size to suggest a slightly larger extinct variety may also have been present. The extant jaguar, *Panthera mesembrina*, is also present, but rare, in the Rancho La Brea deposits. The fossil bones are slightly larger than those of living representatives.

SABER-TOOTHED CATS

Of all the animals known from Rancho La Brea, the saber-toothed cat, *Smilodon fatalis*, most vividly captures the imagination. It has been named the state fossil of California. *Smilodon* is the second most common large fossil from the asphalt deposits (next to dire wolves). It was the size of the modern African lion, but more robust. Its very powerful limbs indicate that the saber-toothed cat used stealth and ambush rather than speed to capture its prey. Nevertheless, saber-toothed cats could probably attain a speed of 25 to 30 miles per hour in short bursts.

The function of the large upper canine teeth has been debated for decades. Their knife-like shape coupled with powerful muscles for lowering the head have led many scientists to believe the canines were used to stab or slash large prey animals. However, other researchers have suggested instead that the saber-toothed cat may have used its canines to bite open the soft underbelly of its prey.

Many of the saber-toothed cat bones recovered from Rancho La Brea show evidence of arthritis and other bone disease. Some paleontologists believe that the saber-toothed cats were social animals, living and hunting in packs that provided food for old and infirm members.

Saber-toothed cats

had powerful limbs to ambush prey, but not even they could leap free from the tar.

Favorite food?
Just meat: bison, sloths, horses, and camels

Best weapon?
Do you need to ask? Four-inch fangs and strong front legs

Did you know?
The saber-toothed cat is California's state fossil.

SMALL MAMMALS

The early excavators at Rancho La Brea were primarily interested in the large extinct animals and collected very few of the smaller remains. Even so, much of the information about the small fossil mammals comes from research on the earlier collections. Most of the species of small mammals found in Rancho La Brea deposits are still living today, and many might still be found in the vicinity were it not for the urban expansion that has taken place during the past half century.

Insectivores are represented by the ornate shrew, *Sorex ornatus*, and the desert shrew, *Noriosorex crawfordi*. Both are known from several specimens. The broad-handed mole, *Scapanus latimanus*, is known from a limb bone and pelvis. Insectivores are among the most primitive of living mammals, and their ancestry can be traced back over 70 million years into the Cretaceous Period.

Two species of bat have been recorded from the asphalt deposits. The hoary bat, *Lasiurus cinereus*, is known only from the end of one limb bone. A smaller, as yet undetermined, species is also represented by one limb bone. Bats had evolved by the early Eocene Epoch, but their delicate bones are seldom found as fossils.

Rodents are much more common in Rancho La Brea material than bats or insectivores. Represented species include Beechey's ground squirrel, *Spermophilus beecheyi*, the Pacific kangaroo rat, *Dipodomys agilis*, the California pocket mouse, *Perognathus californicus*, the valley pocket gopher, *Thomomys bottae*, the California meadow mouse, *Microtus californicus*, the southern grasshopper mouse, *Onychomys torridus*, the western harvest mouse, *Reithrodontomys megalotis*, a packrat of the genus *Neotoma*, and a deer mouse, *Peromyscus imperfectus*.

The origin of the rodents can be traced to fossils from the Paleocene of Asia. Rodents are most closely related to the rabbits and pikas (order Lagomorpha). The squirrels are a specialized offshoot from the most primitive rodent group. In contrast, rats and mice are usually considered the most advanced types of rodents.

The fossil remains of three types of rabbits have been recovered from Rancho La Brea: the brush rabbit, *Sylvilagus bachmani*, the desert cottontail, *Sylvilagus audubonii*, and the black-tailed jackrabbit, *Lepus californicus*. Rabbits and hares belong to an order (Lagomorpha) that is distantly related to the rodents. The modern forms, including those from Rancho La Brea, have descended from a stock first known from Oligocene deposits of Eurasia.

Black-tailed jackrabbit, *Lepus californicus*; fossil skull

ARTIFACTS provide a glimpse into the often elaborate ornamentation worn by Native Americans in this region.

Ornaments

Men would sometimes part their hair in the middle, braid it in the back and then double the braids upward securing them to their head with a hairpin like the one shown above (left).

The pendant (right) was likely strung on cordage and worn around the neck. Jewelry like this may have been worn to show the social status of the individual.

Over one hundred artifacts recovered from Rancho La Brea tell the human story of the tar pits!

Many of these artifacts provide an image of the clothing and adornment worn by Native Americans in the Los Angeles Basin. Small seashells, the most common artifact recovered, may have been brought to the site for trade purposes, but were also used for jewelry and adornment of items, such as bowls. Artifacts made of bone and wood include pendants and hairpins, worn for both every day attire and ritual ceremonies.

Other artifacts point to the domestic and subsistence regimes of the people living in the region. Hide scrapers made of elk horn from the southern San Joaquin Valley were important for creation of clothing and hint at trade networks with northern neighbors. Wooden spear tips and stone point manufacturing material were central to creating weapons that could help capture animals for food. Manos were used in the preparation of both animal and plants resources, and large sea shells would have been used for a variety of things including as bowls, scoops, and containers for the asphalt (tar), which was used as an adhesive and sealant. All of these items may have been similar to ones used by the La Brea Woman. Her remains, which date to approximately 10,250 years old, are the only human remains that have been found in Rancho La Brea deposits.

Birds

The number of bird bones recovered from Rancho La Brea is particularly noteworthy. The bones of birds, like those of most small animals, are rare in fossil deposits because their delicate structure reduces chances for preservation.

At Rancho La Brea, however, the asphalt provided a protective covering for the bones of birds that were trapped and died. The deposits have yielded the largest collection of fossil birds from anywhere in the world, totaling nearly two hundred fifty thousand specimens. Over one hundred forty bird species, twenty-five of which are extinct, have been identified.

Many of the birds represent carnivorous and scavenging species that might have been killed or inadvertently trapped in the asphalt while attempting to feed on other trapped animals. Water birds might have landed in or near asphalt pools, mistaking them for water. Although the fossil birds include a number of these species (e.g., herons, ibises, ducks, geese, plovers, and sandpipers), they make up only a small percentage of the total collection. This suggests that the asphalt seeps were not located in or near permanent standing bodies of water.

EAGLES AND FALCONS

More than twenty species of eagles, hawks, and falcons have been found at Rancho La Brea. The eagles include both species found today in North America: the Golden Eagle, *Aquila chrysaetos*, and the Bald Eagle, *Haliaeetus leucocephalus*. The Golden Eagle is the most common species of bird in the collection, with more than one thousand individuals represented. Also present were extinct species, such as Woodward's Eagle, *Amplibuteo woodwardi*; the Fragile Eagle, *Buteogallus fragilis*; and Grinnell's Crested Eagle; *Spizaetus grinnelli*, whose relatives are now mostly confined to Mexico, and Central and South America. Perhaps the most interesting fossil eagle is *Wetmoregyps daggetti*, whose legs were as long as those of a Great Blue Heron, *Ardea herodias*.

Falcon remains include those of the Prairie Falcon, *Falco mexicanus*; Peregrine Falcon, *Falco peregrinus*; Merlin, *Falco columbarius*; and Kestrel, *Falco sparverius*. Also present was an extinct subspecies of Caracara, *Caracara plancus prelutosus*, a ground-dwelling carrion feeder.

While Rancho La Brea is best known for its extinct mammals, nearly 20 species of extinct birds have been identified from the deposits as well. These include eagles, storks, vultures, geese, and the enormous thirty-pound Teratorns.

Fragile Eagle
Buteogallus fragilis

Grinnell's Crested Eagle
Spizaetus grinnelli

Errant Eagle
Neogyps errans

Woodward's Eagle
Amplibuteo woodwardi

La Brea Caracara
Caracara plancus prelutosus

CONDORS AND VULTURES

The nearly extinct *Gymnogyps californianus*, the California Condor, is the largest living terrestrial bird in North America. What might have been an ancestral form, *Gymnogyps amplus*, was also present in the Rancho La Brea fauna. Although often attributed to a separate species, the ancestral condor can be distinguished from the living form by only minor differences in the skull and by its slightly larger size. Also present was another extinct condor, *Breagyps clarkia*, the Brea Condor, which was slightly smaller than the California Condor and had a longer, more slender beak. A third extinct species of vulture, the Western Black Vulture, *Coragyps occidentalis*, is closely related to the living Black Vulture, *Coragyps atratus*, western North America and much of Central and South America. Remains of the living Turkey Vulture, *Cathartes aura*, are also found in the asphalt deposits.

The condors and vultures presently living in North America are not closely related to the Old World vultures, although both have adopted a scavenging mode of lifestyle. Two Old World vultures are found in the Rancho La Brea deposits, the American Neophron, *Neophrontops americanus*, and the Errant Eagle, *Neogyps errans*.

TERATORNS AND STORKS

The largest birds from the asphalt deposits belong to the extinct family Teratornithidae, whose relationships to other bird families remain unclear. However, they might be related to storks and/or New World vultures. Remains of more than one hundred individuals of Merriam's Teratorn, *Teratornis merriami*, which weighed about 30 pounds and had a wingspan of about 14 feet,

Occidental Vulture, *Coragyps occidentalis*

Brea Condor, *Breagyps clarkia*

La Brea Stork, *Ciconia maltha*

have been recovered. A second, smaller and rarer teratorn, *Cathartornis gracilis*, is also found at Rancho La Brea.

The similarity in appearance between the bones of teratorns and condors led early researchers to believe that both were scavengers. More recent research indicates that teratorns were active predators, stalking their prey on the ground and then swallowing them whole. It is probable that teratorns trapped in the asphalt were seeking their prey around the fringes of the asphalt deposits.

The tallest bird in the Rancho La Brea fauna is a stork, *Ciconia maltha*, which stood about 4½ feet tall. This species is the most common representative of the stork family at this site.

Teratorns

were probably slow flyers because, like condors, they were primarily gliders and depended on winds for soaring.

Best offense?
A long hooked bill used to grab smaller prey.

Favorite food?
Lizards, mice, and other birds

Did you know?
"Teratorn" means "Monster Bird" in Greek.

Merriam's Teratorn, *Teratornis merriami*

TURKEYS

The second most common species of bird found in the asphalt deposits is an extinct species of turkey, *Meleagris californica*, which was slightly smaller than the living Ocellated Turkey of Yucatan, Mexico. Many of the bones of this species are from young birds. Broods of turkey chicks tend to stay together. If one were caught in the asphalt, its cries of distress would attract the remainder of the brood, who might in turn become trapped. Thus, it is possible that the social behavior of the turkeys resulted in entire family groups being trapped at one time.

California Turkey, *Meleagris californica*

OWLS, WOODPECKERS, AND SONGBIRDS

Nine different kinds of owls, including three extinct species, are found at Rancho La Brea. The most abundant species is the Burrowing Owl, *Speotyto cunicularia*, but there are almost as many specimens of the Barn Owl, *Tyto alba*. The Great Horned Owl, *Bubo virginianus*, was also very common. Generally nocturnal hunters, the owls were probably trapped while trying to prey on small animals struggling to free themselves from the asphalt.

Woodpeckers are not common in the Rancho La Brea fauna, but at least seven species were preserved in the asphalt deposits. Most abundant is the Northern Flicker, *Colaptes auratus*.

Many specimens of songbirds, or passerines, are preserved in the Rancho La Brea deposits, but their small bones are extremely difficult to identify with accuracy. Over thirty-five species have been identified so far, the most common being the Yellow-billed Magpie, *Pica nuttalli*; Common Raven, *Corvus corax*; and Western Meadowlark, *Sturnella neglecta*.

This small fossil may or may not be a bird tibiotarsus. Closer inspection and research will be needed to correctly identify what it is.

ENTRAPMENT HAPPENS

Just as they have for thousands of years, many animals are still getting trapped in the asphalt at Rancho La Brea. Insects, birds, squirrels, and lizards, like this southern alligator lizard, are all commonly found stuck in the tar pits today.

Reptiles, Amphibians and Fishes

Because they are often very small, the fossil remains of reptiles, amphibians, and fishes aren't easily detected at the excavation sites.

Most of the smaller vertebrate remains currently known from Rancho La Brea were recovered in the lab during the processing of matrix from Pit 91 and Project 23.

Seven different lizard species have been identified: the desert spiny lizard, *Sceloporus magister*; the side-blotched lizard, *Uta stansburiana*; the coast horned lizard (horny toad), in the *Phrynosoma coronatum* complex; the southern alligator lizard, *Elgaria multicarinata*; the whiptail lizard, *Aspidoscelis tigris*; the western fence lizard, *Sceloporus occidentalis*; and the western skink, *Plestiodon skiltonianus*.

Twelve snake species are known: the racer, *Coluber constrictor*; the gopher snake, *Pituophis catenifer*; the common kingsnake, *Lampropeltis getula*; the glossy snake, *Arizona elegans*; the long-nosed snake, *Rhinocheilus lecontei*; two garter snakes, *Thamnophis sirtalis* and a member of the *Thamnophis couchii* species complex; the striped racer, *Masticophis lateralis*; the ring-necked snake, *Diadophis punctatus*; the night snake, *Hypsiglena torquata* complex; a black-headed snake, *Tantilla* sp., and a rattlesnake of the *Crotalus oreganus* complex.

Abundant remains of the western pond turtle, *Emys marmorata*, have also been found.

Five species of amphibians have been recorded from the asphalt deposits: the western toad, *Bufo boreas*; the arroyo toad, *Bufo californicus*; the California red-legged frog, *Rana draytonii*; a tree frog, *Pseudacris* sp.; and the arboreal salamander, *Aneides lugubris*. Remains of the toads are much more common than those of the other frogs.

All the amphibian and reptile species occur today in Southern California, although not necessarily on the floor of the Los Angeles Basin. The presence of the frogs and the pond turtle indicate seasonal creeks with deeper pools in the area and that the climate was a little more humid than at present.

Remains of three species of fish have been recovered from Rancho La Brea: rainbow trout, *Oncorhynchus mykiss*; arroyo chub, *Gila orcuttii*; and the three-spined stickleback, *Gasterosteus aculeatus*. The stickleback found in the asphalt deposits is less heavily armored than most, similar to an unarmored form now restricted to the upper Santa Clara River. It used to be widespread and abundant in the Los Angeles Basin, as was the arroyo chub. In contrast, rainbow trout are native to coastal drainages from Alaska to Baja California. The trout fossils were less than 5 inches long. When found together, remains of these three fishes indicate that there were permanent slowly flowing streams in the area. The trout thrive in water about 70°F or less; whereas, the stickleback and chub tolerate warmer waters. Cool artesian springs and streams were common before the Los Angeles Basin was desertified by European development.

Only four other freshwater fish species are native to the Los Angeles Basin. Two are lampreys and lack bony elements except for teeth that might be preserved as fossils. The other two fishes, speckled dace, *Rhinichthys osculus*, and Santa Ana sucker, *Catostomus santaanae*, are found in upper stream courses in the San Gabriel Mountains. At first glance, their absence from the asphalt deposits suggests that the ancient streams that flowed through Rancho La Brea were of local origin with headwaters in the Santa Monica Mountains. However, before the early 1800s the Los Angeles River flowed west around the city of Los Angeles and emptied into the ocean via Ballona Marsh. During some periods of the Pleistocene the coastline was several miles farther offshore, and the climate was much cooler and more amenable to the trout and stickleback and possibly to other fishes like salmon.

Lizard jaw (above) and lizard osteoderm (below). Osteoderms are bony deposits that form in the skin, including in the scales of some lizard species.

Invertebrates

Microfossils are sorted under a magnifying lens.

Although fossil vertebrates have received most of the attention from scientists studying the Rancho La Brea fauna, considerable numbers of invertebrate fossils have also been recovered.

These invertebrates, which include more than twenty thousand specimens of mollusks (clams and snails) and over one hundred thousand arthropods (joint-legged animals that include insects and spiders) may, when fully studied, offer much information about environmental conditions in late Pleistocene times.

Freshwater mollusks are represented by at least five species of clams and fifteen species of snails. The aquatic mollusks indicate the presence of ponds or streams through at least part of the year. However, there are at least 11 taxa of terrestrial snails. Interestingly, one species of land snail from the asphalt deposits is not found in California today but only on high rocky slopes in Arizona, New Mexico, and Baja California; the snails of this species found in Pit 91 were probably washed downstream from the Santa Monica Mountains.

Likewise, shells of marine mollusks, several of which are found in the asphalt, may have been deposited by birds or introduced as human artifacts.

Seven different orders of insects have been found—grasshoppers and crickets, termites, true bugs, leafhoppers, beetles (the most abundant because their hard bodies preserve well), flies, and ants and wasps. Other arthropods are represented by millipedes, scorpions, several families of spiders, ostracods ("seed shrimp") and isopods (sow bugs or pill bugs).

Some of the fossilized insects, especially the bugs and beetles, were aquatic and would have been preserved naturally in stream or pond deposits. However, more than half of the represented species lived on land and are only infrequently found as fossils. Some of the beetles and flies were carrion feeders and may have become trapped while feeding on carcasses of animals that were stuck in asphalt; the other kinds of insects were probably trapped when they were blown into, or inadvertently crawled over, the sticky asphalt.

Although early workers on Rancho La Brea insects identified many extinct species, modern studies indicate that nearly all the fossilized specimens represent insects that are still living today (although some, because of climatic change, are no longer found in the Los Angeles area). The only two extinct species now recognized from Rancho La Brea are dung beetles. These beetles are dependent on plentiful supplies of dung for their life cycle, and their extinction might have coincided with that of the megafauna mammals at the end of the Pleistocene.

Mexican rams-horn snail, *Planorbella tenuis*

Ground beetle, *Carabidae* sp.

Plants

Fossil seeds, Manzanita, *Arctostaphylos* sp.
Habitat: Chaparral

Purple sage, *Salvia leucophylla*
Habitat: Coastal Sage Scrub and Chaparral

Fossil cone, Monterey cypress, *Hesperocyparis macrocarpa*
Habitat: Coastal Sage Scrub

Fossil seeds, California sycamore, *Platanus racemosa*
Habitat: Riparian Woodland

Fossil plant material from Rancho La Brea consists of wood, leaves, cones, seeds, and microscopic remains of diatoms.

Well over a hundred thousand plant fossils have been recovered. Some of the material is from plants that lived in the immediate area, and some material shows evidence of being transported by floodwaters or streams. Studies of the plant material and late Pleistocene landscape have suggested that four plant associations were present in the area—chaparral, mixed evergreen/redwood forest, riparian woodland, and coastal sage scrub.

CHAPARRAL

Chaparral can be characterized as being composed of tall (4 to 10 feet), usually densely packed, deeply rooted, woody bushes that are dependent on fire for vitality. These often impenetrable associations typically occur on steep, rocky slopes. Although probably not much different in appearance than today, the chaparral of Pleistocene times included many plants that no longer occur in the Los Angeles region. The dominant plants of this association were chamise, California lilac, scrub oak, manzanita, walnut, elderberry, coffeeberry, and poison oak. Juniper and foothill pine were probably scattered in the chaparral in more open, drier areas. Coast live oak probably occurred in groves on north-facing slopes, in smaller canyons, and on the lower slopes of deeper canyons.

MIXED EVERGREEN/REDWOOD FOREST

Coast redwood, California bay, and dogwood, species usually associated with mixed evergreen and redwood forests, occurred in deep protected canyons. These plants probably grew in proximity to coast live oak and stream bank plants.

RIPARIAN WOODLAND

The plants that inhabited stream margins and springs form a third association called a riparian woodland. This group included sycamore, alder, arroyo willow, dogwood, blackberry, poison oak, and numerous herbs, such as sedges, bulrushes, and mugwort. Where the streams crossed the plain, the association consisted of arroyo willow, red cedar, occasional coast live oak, sycamore, elderberry, walnut, numerous herbs, and possibly Bishop pine.

COASTAL SAGE SCRUB

The typical coastal sage scrub association is composed primarily of 3–4 foot tall, drought-deciduous (losing their leaves during the dry summer months), woody bushes that are interspersed with herbs and grasses. In Pleistocene times, this association, punctuated by valley oak and groves of close-cone pine, covered the plain. The dominant plants of this association included California sagebrush, black sage, and buckwheat. Manzanita and juniper also occurred frequently in this association. Valley oaks were scattered at higher elevations on the alluvial fans. Groves of Monterey cypress and Monterey pine occurred in favorable sites on the plain.

Creek Dogwood, *Cornus sericea*
Habitat: Mixed Evergreen/
Redwood Forest

Lemonade berry, *Rhus integrifolia*
Habitat: Coastal Sage Scrub and Chaparral

Western elderberry, *Sambucus nigra* subsp. *caerulea*
Habitat: Coastal Sage Scrub and Riparian Woodland

Salt or Quail bush, *Atriplex lentiformis*
Habitat: Coastal Sage Scrub

Oregon grape, *Berberis aquifolium*
Habitat: Mixed Evergreen/Redwood Forest

Purple sage, *Salvia leucophylla*
Habitat: Coastal Sage Scrub and Chaparral

White sage, *Salvia apiana*
Habitat: Coastal Sage Scrub and Chaparral

Coffeeberry,
Frangula (Rhamnus) californica
Habitat: Coastal Sage Scrub and Chaparral

St. Catherine's lace,
Eriogonum giganteum
Habitat: Coastal Sage Scrub and Chaparral

California sycamore,
Platanus racemosa
Habitat: Riparian Woodland

Arroyo willow, *Salix lasiolepis*
Habitat: Riparian Woodland

California sagebrush, *Artemisia californica*
Habitat: Coastal Sage Scrub and Chaparral

Plants from the Ice Age

Long before palm trees lined its busy streets, Los Angeles was an oasis of pine, sage and buckwheat. Scientists at the La Brea Tar Pits Museum have recreated this original habitat with the Pleistocene Garden, a prehistorical landscape in Hancock Park representing the native vegetation of the Los Angeles Basin 10,000 to 40,000 years ago. Planned entirely from a plant list that was gathered from 35 years of research in the Pit 91 fossil excavation, the garden was started in 2004 and was organized by vegetation types: Coastal Sage Scrub, Riparian Woodland, and Mixed Evergreen/Redwood Forest. A fourth type, Chaparral, was added in 2010. Whether you want to read its informative signs about native plants, gaze at its soothing stream or just take in its many fragrances, the Pleistocene Garden is a welcome step back in time, away from the hustle and bustle of Wilshire.

The Future

The study of the Rancho La Brea fossils is still far from complete.

For more than one hundred years, the vast majority of research at Rancho La Brea focused on the large mammals—mammoths and mastodons, saber-toothed cats and dire wolves, bison, camels, lions, horses, and ground sloths. Only recently have scientists begun to realize the important contributions that microfossils can make to our understanding of past ecosystems, environments, and climates. As scientists continue to research the tar pits vast treasure trove of small vertebrates, invertebrates, and plants, more species will be added to our list of known plants and animals from the tar pits at Rancho La Brea, and more will be learned about Los Angeles' past, as well as its future.

At present, we know that the Rancho La Brea fossils cover a time period spanning from over 50,000 years ago to the present. As you walk around Hancock Park today, you can observe active asphalt seeps capturing leaves, insects, and occasionally even small vertebrates, just as they have done for tens of thousands of years. Continued radiocarbon dating and resolution of the exact age of each concentration of fossils will provide further information about ecological and evolutionary changes that took place during this interval of time.

The Fossil Lab, located inside the museum, is a glass-walled paleontological laboratory. The lab offers visitors a behind-the-scenes opportunity to observe as fossil specimens are cleaned, examined, and cataloged.

Pit 91 is currently 15 feet deep and it is estimated that the fossil deposit extends another 3 to 8 feet further below ground.

Continued investigations of the ways in which the deposits were formed will help us understand how the fossils came to be preserved here, and perhaps, why there are so many examples of some animals but so few of others.

THE IMPORTANCE OF THE FOSSILS

The Rancho La Brea fossil deposits provide a spectacularly comprehensive record of life in the Los Angeles region toward the end of the Pleistocene epoch. The preservation of many of the specimens is unusually complete, permitting detailed studies of the form and function of animals that are long extinct. Exceptional asphaltic preservation of collagen and chitin permits scientists to radiocarbon date bones, plants, and insects, placing each specimen precisely along a 50,000 year timeline. Plants and small animals, including reptiles, insects, and mollusks, give clues to changes in the climate of ancient Los Angeles—changes that may be intimately tied to the fate of the iconic large megamammals that typify the Rancho La Brea biota. Tooth wear, stable isotopes, and even plant fragments preserved in the dental plaque of some herbivores provide insight into what these animals ate, and how their diets changed in relation to changing climate and environmental conditions over thousands of years.

The importance of the fossils recovered from Rancho La Brea, and the deposits that remain in and around Hancock Park, cannot be overestimated. They hold the key to many current and future questions about the effects of climate change, human activities, and extinctions on animals and ecosystems—questions that are more pressing today than ever before.

Museum paleontologists bring the Ice Age back to life for students.

The Museum Collections

The vast collections in the Museum at Rancho La Brea are regarded as the world's richest treasure of Ice Age fossils.

The Rancho La Brea collections are conservatively estimated at more than 3.5 million specimens representing over six hundred species of animals and plants, and are the basis for our knowledge of life in North America in the late Pleistocene and Holocene. The collections also include geological samples, archaeological artifacts, historical objects, art objects, a library and archive, and collections of correspondence and other ephemera related to Rancho La Brea.

The fossil collections are core resources for research into the late Pleistocene of North America. Museum staff, research associates, visiting scientists, professional paleontologists, and students from all over the world frequent the collections year round. Today's research ranges from molecular analyses on fossils to traditional taxonomic studies of extinct fauna to investigations of the geochemistry and living biota of the modern asphalt seeps. Increasingly, research projects seek to leverage Rancho La Brea's remarkably rich collections to investigate the nature of past species and ecosystems, and how they changed over time in response to climate change and other factors. After more than a century of study, Rancho La Brea's greatest and most important contributions to our scientific understanding of our world are almost certainly yet to come.

Research and Collections staff curate over 3.5 million specimens, including these Harlan's giant ground sloth toe bones.

The Observation Pit was Rancho La Brea's first on-site museum. It connects visitors to the past, present, and future of paleontology in Los Angeles.